Hilmi Armoush

Business Model Innovation For Shipbrokers In The Drybulk Industry

Hilmi Armoush

Business Model Innovation For Shipbrokers In The Drybulk Industry

A strategic insight into a new shipbroker's role: Innovation Intermediary

LAP LAMBERT Academic Publishing

Impressum / Imprint
Bibliografische Information der Deutschen Nationalbibliothek: Die Deutsche Nationalbibliothek verzeichnet diese Publikation in der Deutschen Nationalbibliografie; detaillierte bibliografische Daten sind im Internet über http://dnb.d-nb.de abrufbar.
Alle in diesem Buch genannten Marken und Produktnamen unterliegen warenzeichen-, marken- oder patentrechtlichem Schutz bzw. sind Warenzeichen oder eingetragene Warenzeichen der jeweiligen Inhaber. Die Wiedergabe von Marken, Produktnamen, Gebrauchsnamen, Handelsnamen, Warenbezeichnungen u.s.w. in diesem Werk berechtigt auch ohne besondere Kennzeichnung nicht zu der Annahme, dass solche Namen im Sinne der Warenzeichen- und Markenschutzgesetzgebung als frei zu betrachten wären und daher von jedermann benutzt werden dürften.

Bibliographic information published by the Deutsche Nationalbibliothek: The Deutsche Nationalbibliothek lists this publication in the Deutsche Nationalbibliografie; detailed bibliographic data are available in the Internet at http://dnb.d-nb.de.
Any brand names and product names mentioned in this book are subject to trademark, brand or patent protection and are trademarks or registered trademarks of their respective holders. The use of brand names, product names, common names, trade names, product descriptions etc. even without a particular marking in this works is in no way to be construed to mean that such names may be regarded as unrestricted in respect of trademark and brand protection legislation and could thus be used by anyone.

Coverbild / Cover image: www.ingimage.com

Verlag / Publisher:
LAP LAMBERT Academic Publishing
ist ein Imprint der / is a trademark of
OmniScriptum GmbH & Co. KG
Heinrich-Böcking-Str. 6-8, 66121 Saarbrücken, Deutschland / Germany
Email: info@lap-publishing.com

Herstellung: siehe letzte Seite /
Printed at: see last page
ISBN: 978-3-659-57658-4

Zugl. / Approved by: London, Middlesex University & lloyd's Maritime Academy, Master Thesis., 2012

Copyright © 2014 OmniScriptum GmbH & Co. KG
Alle Rechte vorbehalten. / All rights reserved. Saarbrücken 2014

Abstract

Purpose – The purpose of the paper is to provide strategic insight for shipbrokers on how to increase customer value proposition which meet with perceived value for their target customers and sustain a strategic role within supply networks through the business model innovation approach.

Design/methodology/approach – The paper aims to identify the paradigm competed-on attributes among shipbrokers within the drybulk industry and test how such attributes are perceived by drybulk shipowners. The strategy canvas was used to identify the differences and similarities within different shipbroking strategic groups and find out on the paradigm competed-on value attributes and strategies. The paper then analyzes the gaps of such attributes with the perceived value from shipowners as asset owners and ship operators.

Findings – Most shipbrokers within the drybulk industry follow similar strategies in terms markets covered, target segment, customer value proposition and the number of brokers needed to pursue their strategies. Most players within the shipbroking industry also focus on the personal relation and brand reputation as main differentiated attributes. While most modern ships work within the four shipping markets: freight, sales and purchase, newbuildings and demolition – shipbrokers are focusing their efforts on covering the freight and the sale and purchase markets while targeting charterers/cargo traders. The main reason for such approach is perhaps the paradigm role of shipbrokers as defined by the Baltic Exchange. When such attributes were compared with the perceived value of our sample of shipowners, many gaps were identified which show that the industry is over or under investing in many attributes that may not lead to sustainable growth and thus increase substitute threats. Therefore, the shipbroking industry needs to be more innovative in order to sustain their role within supply networks and increase their value offerings to customers. Innovation intermediary role was recommended as a start for redefining the role. Companies need to apply the business model innovation approach to sideline competition and open new market spaces for a sustainable competitive advantage.

Research limitations/implications – Due to the limited resources and time, data collection could only take place during the month of August 2011. It is recommended for future studies to do a much wider surveys from wider samples to analyse the paradigm strategies and perceived value from most industry stakeholders. It is also recommended to include other attributes for better comparison such as technologies used, cultures, systems and structures.

Keywords – Business Model Innovation, Value Innovation, Shipbrokers, Shipping Markets, Customer Perceived Value.

Table of Contents

Abstract 1

Background 4

Literature Review 6

   The Shipbroker Role: 6

   The competitive environment: 10

   Value Innovation: 12

   Business Model Innovation 16

Method 20

   Approach 20

   Design 21

   Methodology 24

   Sample Selection 25

   Research Limitation 26

Results 27

Discussion 39

   Shipbroking Paradigm Strategies 39

   Business Model Innovation – A source of competitive advantage 42

   A New Shipbroking Role 44

Conclusion & Recommendations 47

References 49

Background

Shipping is considered as one of the most international industries carrying over 90 percent of global trade – of goods and energy – by sea in a cost efficient, clean and safe manner (IMO2011). The industry has successfully evolved over the years to sustain its strategic role as a central part of globalization and managed to effectively and efficiently integrate with other global supply networks (Stopford 2009).

Since the first cargo was moved by sea over 5000 years ago, the industry has seen many diverse changes in its technical, technological, legal and commercial functions which have transformed it from a loose system run by traders to a highly integrated global industry specialized in the transport of cargo by sea (Stopford 2009). Globalization was a major driver to promote the international flow of trade, people, investment, technology and capital between countries and - as a derived demand - the globalization of the international shipping market has also been accelerated further (UNCTAD 2010).

However, the sustainable shipping initiative claims that over the last two decades, the shipping industry has been seeing major changes on the macro and microeconomic levels. Factors such as economic and political power shifting, changes in trade patterns, slower/uncertain economic growth, fluctuating oil prices as well as severe environmental changes are all major causes that have been shaping future challenges (The sustainable shipping initiative 2011). Challenges that will impact almost every organisation within the shipping industry and will cause a major paradigm shift in how organisations work (Lorange 2009). However, being in a highly mature, volatile and fragmented industry this may impose additional challenges that would affect growth plans and objectives (Lorange 2009). In our modern world, the higher expectation and dynamic buying behavior are becoming more environmentally anxious and supply chain oriented (Kotler and Keller 2010).

In this paper, I argue that shipbrokers must leverage on such emerging opportunities and respond to changing customers' needs and buying behaviour. Being a central part of the industry, shipbrokers must aim to sustain their strategic role within the supply network. In order to do so, they must undergo a fundamental restructuring of the way they operate to create a leap in their customers' value proposition and meet with their dynamic perceived value (Lorange 2009). Instead of pursuing further cost reduction, process reengineering, further segmentation, increased marketing budgets or any other traditional strategy; shipbrokers must play the game differently.

The paper is structured as follows: literature review is presented first, followed by the methodology section. Result and discussion of collected data are presented next which will tend to answer the following questions:

- What value-attributes are commonly offered by shipbrokers and how such attributes are perceived by dry bulk shipowners?
- What is the paradigm shipbroker role within the dry bulk shipping industry and how it can be improved?
- Why do shipbrokers have to develop business model innovation? And how it can be a source of competitive advantage?
- What are the important components in business models to sustain a competitive advantage for managing the future?

Literature Review

The Shipbroker Role:

Working in shipbroking industry has been experiencing continuous changes since the 17th century when traders would drop by the Virginia and Baltick coffee house to trade and draw up agreements for the sale and transportation of goods by sailing ships (the Baltic exchange 2011). At that time, shipbrokers attending the Baltic floor could obtain concessions on price and terms of commodities by meeting face to face which neither buyers nor sellers could manage (The Baltic Exchange 2011).

Throughout the years, the communications revolution introduced many innovative products such as the telex machine, faxes, e-mail and the internet which gradually reduced the importance of the face to face meetings and changed the paradigm competitive environment in the industry (The Baltic Exchange 2011). Moreover, as trade grew – the raising of capital to finance ship acquisitions became easier (the Baltic Exchange 2011). During 18th century the concept of 'sixty-fourth' company allowed investors to buy part of a ship as a standalone investment. Then, towards the second half of the century, the company Act 1862 allowed investors to be protected from liability claims by registering a joint stock limited liability company instead of relying on a shipbroker to raise capital from friends and family and operate ships (Stopford 2009). These factors as well as others have led to contemporary problem for shipbrokers – as owners, agents and traders at that time - and therefore shipbrokers had to shift their role towards being more of an intermediary rather than being an agent (Fischer 1993).

In 1877, the shareholders for the 396-ton barquentine Julia Fisher signed an agreement with an individual named George Bell to sell their craft. The contract empowered Bell to sell the barquentine "at any place outside the Dominion of Canada for a period of twelve months" at a price not less than £2500. During negotiations with potential buyers, Bell described himself in shipping documents variously as a "merchant", "shipowner", "commission agent", and "shipping agent". But when he signed the agreement between parties he designated yet another occupation—"shipbroker" (Fischer 1993). While this transaction was not in itself of any special historical significance, it could be argued that it had an innovative proposition that added tremendous value to shipowners allowing them to offer and sell their ships to international buyers (Fischer 1993). Brokers like George Bell increasingly became essential to the shipbroking industry and its international intermediary role.

During the past few decades, the web technology was further developed to do almost anything (Kotler and Keller 2010). It was increasingly used by many industries as a main tool for communication, gathering and putting together databases as well as creating extra added-value services for international customers in a cost and time efficient manner (Kotler and Keller 2010). The IT sector had major effect on the shipbroker role as more and more shipbrokers started to rely on computers to do much of their work (Prasad 2008). In his article (changing role of shipbrokers 2008) Mr. Krishna Prasad argues that due to the evolution in trade specialization in shipping and the impact of modern communication; the role of shipbrokers became more pronounced. According to his findings, the modern shipbroker should to a large extent act as a collector, judge and distributor of information. He proposed a new definition of the shipbroker: *"an information network; a network of people as well as technology that facilitates information exchanges"*. He claims that the changing roles of the modern shipbrokers should be more of information intermediary to include:

- Finding ships for cargoes and cargoes for ships
- Source of information
- Market trends and an adviser to owners and charterers
- A dispute resolver

From the legal point of view, such role may imply many legal problems for shipbrokers. According to an article listed in the Lloyd's list titled (English Law has trouble finding definition of a shipbroker 2006), The article expressed that unlike other profession roles such as Estate Agents, there is no definition that can be drawn form an Act of parliament to cover shipbrokers. According to Mr. Jaimson's legal opinion, a sale and purchase broker could be defined as being a *"professional representative acting on behalf of a party engaged in negotiations for the buying and selling of as ship"*. Such definition raises two sources of duties and liabilities. First, the relationship between the shipbroker and his client is a contractual one. Secondly, there are duties and liabilities arising from the services the broker provides as English law imposes additional duties on those act for others (Lloyds List 2006). A common source of claims is brokers relying on what can be found in their databases and circulating details or information without checking with their correctness, even if the information was received from another broker, the shipbroker will still be liable and forced to seek an indemnity down the line (Lloyds List 2006).

In practice however, the modern shipbroker mainly competes for being the first to advise his clients of business opportunity and successfully executes the negotiations and contractual terms between parties. Shipbrokers' basic function is to bring two parties together and negotiate a charterparty or sale agreement. Once agreement is reached and contracts signed, the broker is entitled for the agreed commission (Branch 1995). Although a shipbroker may act as a sole broker for one or multiple principals, they will be in most cases involved in functions such circulation of tonnage and business to potential clients, negotiating main terms, finalizing details, post fixtures etc (the virtual shipbroker 2011).

Accordingly, it could be seen that the current shipbroker role has evolved from first as 'Agent' to 'International Intermediary' to 'Information Intermediary' and is now increasingly becoming 'Execute Only Intermediary'.

The shipbroker is currently defined by the Baltic Exchange as *"someone who arranges the ocean transport of goods and commodities by sea, the employment of a vessel or buys and sells ships on behalf of his clients"*; it describes their role as *"intermediaries between shipowners and charterers or the buyers and sellers of ships"* (Baltic Exchange 2011). According to this definition and role, shipbrokers could be seen as intermediaries working within the freight and the sale and purchase markets. Martin Stopford (2009) identified four main closely related shipping markets: new buildings, freight, sale and purchase and demolition (Stopford 2009). He claimed that each market has its own characteristics and players but they all work together linked by industry cashflow. Most modern shipowners trade in all of the four markets. Depending on their fundamental strategies and sentiments; cash can be transferred between shipowners' bank accounts and the different players within the four markets (Stopford 2009).

The important role of the Baltic Exchange-London as well as the creation of key broking clusters in cities such as London, Oslo, Piraeus, New York, Singapore etc have so far helped shipbrokers over the years to evolve and stay relevant despite the threat to their traditional intermediary role. However, the initiation of freight derivatives trading and the shipping boom in 2004-2008 attracted new specialized market players such as investment banks and hedge funds into the shipbroking industry which further increased competition (the maritime advocate 2011). This comes in tandem with other challenges such as supply/demand imbalance, accelerated technological advances, diffused knowledge, and product commoditization (Lorange 2009). Those factors as well as others may have substantial influence on the shipbroker role and its sustainable position, the main questions remain: Is the current definition of shipbroker role really covers the existing shipbroking function? And, will such definition allow shipbrokers to maintain a sustainable strategic role within the upcoming challenges and increasing competition? Or, do they have to reconsider and redefine their role in order to stay relevant and valuable?

The competitive environment:

Today, a sheer number of shipbrokers operate within the four shipping markets. The size of shipbroking companies vary from small boutiques serving niche segments to large corporations operating through multiple divisions and business units with full-market coverage across six continents (the virtual shipbroker 2011).

The continuous market growth, low capital requirement and diffused technical and technological know-how are inviting more and more firms to enter the market to compete with similar locked-in strategies, systems, structures, cultures and paradigms (Grant 2008). Such situation is adding the pressure on leading firms as brand switching is becoming a norm by many customers due to undifferentiated commodity-based perceived offerings within relatively highly fragmented industry. What makes the situation even more challenging is that more customers do not perceive any value from shipbrokers' offerings and are increasingly working direct with their end-customers through less distribution layers (Lorange 2009).

The sustainable shipping initiative claims that over the past two decades, the shipping industry has been experiencing major changes driven by multiple macro and microeconomic changes. Factors such as economic and political power shifting, changes in trade patterns, slower/uncertain economic growth, fluctuating oil prices as well as severe environmental changes are all major causes that have been shaping future challenges for all players (The sustainable shipping initiative 2011). This means that the shipping industry is preparing for a paradigm shift that will ultimately affect success or failure for most players (Lorange 2009). The critical task for shipbroking firms therefore remains to constantly meet changing demand of their customers and sustain their strategic role within global supply network. This may call for fundamental restructuring of the way they operate and develop new strategies to meet with changing buying behaviour and customer needs (Lorange 2009).

One excellent way to help understand, predict and manage change is to analyze the industry life-cycle and meet critical success factors for each stage (Grant 2008). Similar to product life-cycles, industries go through four stages: introduction, growth, maturity, and decline. With each stage, market characteristics and competitive environment change and so the critical success factors. According to Utterback and Abernathy model (1975), firms within new industries compete mainly on product differentiation and profitability - as markets mature and customers' needs become more defined - firms shift their focus to competing on cost, economies of scale and more heavily on resources and capabilities in order to offer more specialized and efficient processes (Grant 2008). The idea concerning the role of resources and capabilities as a source of competitive advantage has been developed further by scholars and executives to become known as the resource-based view of the firm (J.B Barney 1991). C.K.Prahalad and Gary Hamel (1990) have explained how management's ability to consolidate capabilities and resources into core competencies could be a powerful source of competitive advantage and an effective way to adapt quickly to changing opportunities.

In order for firms to develop competitive advantage within the shipbroking industry and attract more customers they must identify and meet their key success factors. This could be achieved by meeting two criteria: first, they must supply what customers want to buy; second, they must survive the competition. (Grant 2008). Hagel and Singer (1999) described three types of specialist firms, each focusing on a different aspect of the value chain – customer relations, product innovation, and infrastructure. They argued that each type should focus on different competencies and success factors critical to offer added value propositions to their target markets. Building on their findings, Dr Lorange (2009) provided in his book (Shipping Strategy) four types of shipping organisations with each requiring different success factors and competencies that collectively contribute to the industry's value chain – owning steel, using steel, operating steel and innovating around steel (Shipping Strategy 2008). Dr Lorange claims that the era of the fully integrated company undertaking various functions such as buying and selling ships, financial operations and in-house crewing and operation is largely past and the future trends is towards decomposing the value chain through vertical integration while focusing on core competencies. This means that to success, companies must move to specialization and concentrate on delivering what customers want. It is the overall industry's value chain that matters, not only one or small part of the industry (Lorange 2009).

However, in mature industries such as shipping, it has become increasingly apparent that most core competencies are being easily replicated by competitors and what used to be a source of competitive advantage becomes an industry diffused standard (Grant 2008). As companies become experts in their core competencies, they continue on developing differentiated increased value through finer segmentation and continuous process reengineering. Firms also continue to refine their strategic objectives and metrics to ensure more effective execution and control (Johnson, Christensen and Kagermann 2008). But there comes a time when those competencies reach the upper thresholds and when markets become fully mature (Johnson 2010). Companies will then need something more fundamental than gradual growth: they will need to deliver new value. Here when Mark W. Johnson (2010) calls for the time to innovate something *"more core than the core"* which is business model innovation.

Value Innovation:

The current shipbroking competitive environment is described by Kim and Mauborgne (2005) as Red Ocean. Here industry boundaries are defined and accepted and the competitive rules of the game are known. As the market space gets crowded, prospects for profits and growth are reduced, products become commodities, and cutthroat competition turns the red ocean bloody. They argue that if companies wish to find high, profitable growth they must go beyond red ocean strategy and aim for what they call the Blue Ocean strategy. They describe blue oceans as: *"uncontested market spaces created by companies to avoid head-to-head competitors and create value to customers"* (Kim and Mauborgne 2005).

At the heart of the blue ocean strategy is finding new value for new markets. Kim and Mauborgne (2005) provide four guiding principles for the successful formulation of blue ocean strategy. The first is to reconstruct market boundaries and create uncontested market space. This could be achieved by looking across six paths which focus on looking across alternative industries, across strategic groups, across buyer groups, across complementary product and service offerings, across functional-emotional orientation and even across time. This gives companies keen insight to reconstruct market boundaries and create blue oceans. The second approach termed as 'focus on the big picture, not the numbers'. This approach tackles possible planning risks through designing strategic planning process to create value innovations away from incremental improvements. The third approach – reach beyond existing demand - not through finer segmentation but by building on commonalities across non-customers and other groups. The fourth approach provides the strategic sequence to provide a leap in value as well as build a viable business model to produce and sustain profitability.

Kim and Mauborgne (2005) used an excellent analytical tool called the strategy canvas. Unlike other tools currently in common use, the strategy canvas is a diagnostic and an action framework for building blue ocean strategy. Its' value curve helps capture the current state of play in the known market space and analyze what customers receive from existing competitive paradigms among different strategic groups within same industry. In order to build a successful blue ocean strategy, they proposed the four action framework grid – reduce, raise, create and eliminate – which can be used to break the trade-offs between differentiation and low cost and to create a new value curve.

Several scholars have defined and provided a typology for classifying value and its composition. For example, Anderson and James A. Narus (1998) define value as the *"worth in monetary terms of the technical, economic, service, and social benefits a customer company receives in exchange for the price it pays for a market offering"*. Kotler and Keller (2010) are in agreement with this but recommends that in order to explore, develop, communicate and deliver superior value; companies should fully understand their customers' motivational and psychological cues and values and design market offerings that meet with what they call "customer-perceived value". They defined customer perceived value as *"the difference between the prospective customer's evaluation of all the benefits and all the costs of an offering and the perceived alternatives"* (p.381). Based on the discussion above, the value for a product (or service) can be broadly expressed as the differences between total benefits and total costs, which can be expressed by the following equation: **customer perceived value = total benefits – total costs.** Kotler and Keller (2010) illustrate what could be considered as total benefits and total costs as shown in figure 1.

Figure-1: Customer Perceived Value

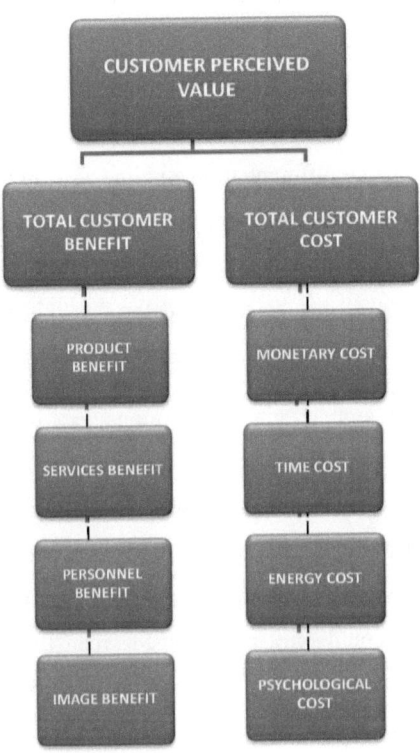

Source: Kotler & Keller (2010) P.381

Porter (1985) discusses how the value chain is used as a tool to deliver value. According to his model, every firm is a synthesis of activities performed to deliver its final market offering. Based on their positioning strategy, firms should properly align and integrate their value chain management to deliver value for their target markets. However, in mature and highly fragmented industries such as dry bulk shipbroking, delivering low cost or differentiated value to customers has never been an easy and straightforward task. Kim and Mauborgne (2005) suggested that companies need to reject the fundamental view of conventional strategies that a trade-off has to be made between value and cost and pursue what they call "value innovation". Unlike 'strategy' which is more concerned with competitive positioning and beating competition, 'value innovation' focuses on making the competition irrelevant by creating a leap in value which allow companies to pursue differentiation and low cost simultaneously. (Kim and Mauborgne 2005) By identifying and analysing the industry value curve, companies can achieve cost savings through eliminating and reducing factors the industry mostly competes on, while buyer value can be achieved through raising and creating value attributes that best meet with customers' perceived value (Kim and Mauborgne 2005).

The literature focuses on how strategy innovation brings great benefits to firms through changing the rules of the game, sideline competition, exploring untapped market spaces and offering higher value to new markets. However, discovering new value to new markets is only one part of the value innovation. It is the whole-system approach that makes the creation of blue oceans a sustainable strategy. Kim and Mauborgne (2005) claim that value innovation could only be achieved when the whole system of the company's utility, price and cost activities are properly aligned. Joan Magretta (2002) claims that strategic innovation could be achieved through a good business model which tells a story on how the firm works, he argues that: *"while strategy deals only with competition; business model could be a source of competitive advantage being a powerful tool that focuses on how elements of the system fit into a working whole to create and deliver value to target customers"*.

Business Model Innovation

By looking across other industries such as no-frill airlines, department stores, entertainment, and online brokerage, it is evident that the business model approach was a main tool that separated winners form losers (Johnson 2010). It didn't matter whether firms were small or large, whether they operated in attractive or unattractive industries, whether they were new entrants or established incumbents, what mattered is their ability to pursue strategy innovation through the business model innovation approach (Johnson 2010). Such firms did not try to play the game better than competitors but rather they managed to avoid head-to-head competition by actively adopting a different business model that aim to change the rules of the game in the industry (Markides 2008).

The term "business model" has been a managerial buzzword since the internet burst with an increasing number of books and articles focusing on this issue (Lehmann-Ortega and Marc Schoettl 2008). The concept of business model innovation was described by Christensen and Raynor (2003) as "disruptive innovation", Kim and Mauborgne (2005) called it "the right strategic sequence" while Mark W. Johnson (2010) called it "seizing the white space".

It is an excellent tool for value innovation (Johnson 2010). According to Markides (2008) the business model innovation is: *"the discovery of a different and difficult to-imitate business model in an existing industry. The innovation enlarges the market by attracting noncustomers to the product or service or by encouraging existing consumers to consume more"*. Mark Johnson (2010) defines a business model as: *"a representation of how a business creates and delivers value, both for the customer and the company"*. In order for companies to create a powerful business model, he proposed the beautiful four-box framework as shown in figure-2.

Figure-2 – The Four-Box Business Model Framework

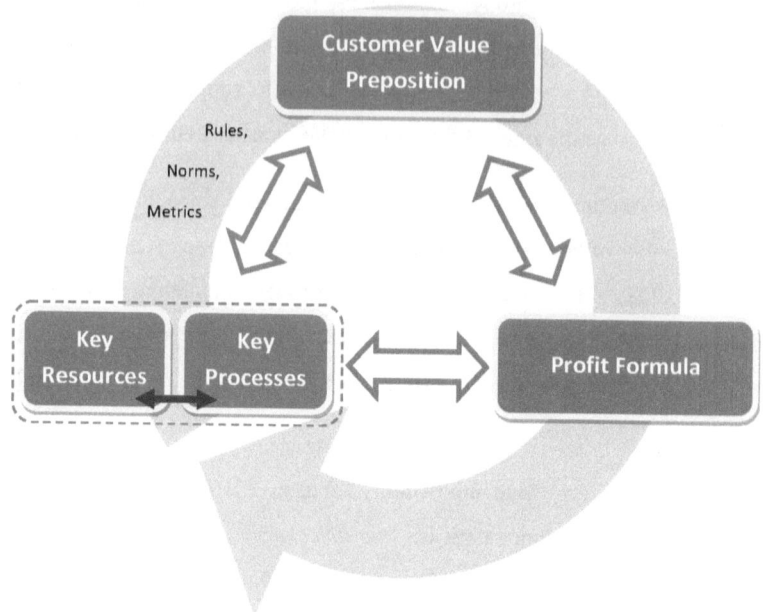

Source: Mark W. Johnson 2010 P.45

The first box is the customer value proposition. It could include a product, service or combination thereof that helps customer do more effectively, conveniently, or affordably a job that they have been trying to do (Johnson 2010). Johnson (2010) argues that a powerful customer value proposition is the keystone of all successful business models which can only be achieved by developing a comprehensive understanding of customers' job-to-be-done or perceived as valuable and develops an offering that does better than alternatives at lower prices (Johnson 2010).

The second box is the profit formula; this explains how the company will create value for itself and shareholders (Johnson 2010). This could be achieved by innovatively reconsidering and redefining the following four interrelated variables: the revenue model, cost structure, target unit margin and resource velocity (Johnson 2010). The interaction between those variables must always support the customer value proposition in a profitable way. The third and fourth elements of the business model, key resources and key processes are the means by which the company delivers the value to the customer and itself (Johnson 2010). They include the critical assets, skills, activities and routines that enable companies to fulfill the customer value proposition and the profit formula in a repeatable, scalable fashion. Johnson (2010) argues that the four interdependent business model elements when properly aligned and integrated, they should provide a consistent and profitable business. A change to anyone of the four affects all the others and the system as a whole (Johnson 2010).

However, business model innovators should not only aim to discover new products or services in a profitable way, but must also aim to redefine the business they are in, redefine the customers and the products or services offered to them and redefine how they will be provided (Markides 2008). Drawing from Abell's work (1980), Markides (2008) identified four main questions which the sums of their answers form an innovative business model:

- What business I am in?
- Who my customers really are?
- What to offer them?
- How to play the game?

Critical to answer the four questions, companies must aim to break the rules of the game by redefining the existing paradigm in terms of role, segment, value proposition as well as core competencies and corporate environment that support such innovation (Markides 2008).

Method

Approach

The aim of this paper is to find out how shipbroking offerings are perceived by dry bulk shipowners and provide strategic insight on how to offer increased customer value proposition through business model innovation approach. As an explanatory research; the paper adopts a functionalist paradigm philosophy to meet its objectives as follows:

- Select appropriate models and theories for exploring, evaluating and innovating effective value attributes.
- Develop a selective network of shipbrokers and shipowners from which primary data will be collected.
- Analyze and interpret the data to clarify monetary, economic, motivational and psychological perceived value offered by shipbrokers as perceived by sample of dry bulk shipowners.
- Test relationships with hypotheses to analyze gaps and highlight areas of increased value.
- Make future recommendations on the various hard and soft issues critical to sustain an innovative business model within specific shipbroking context.

Shipbrokers will benefit from the systematic and practical methodologies used in this paper to analyze their own strategic positioning compared to other strategic groups within the shipbroking industry and scale their market offerings as perceived by selected sample of ship owners. They will also benefit from the strategic insights that are discussed in this paper and how to redefine their role and apply new rules of the game through the business model innovation approach. While this paper discusses on ways to develop business models and create a leap in value, it does not however aim to explore or create a new business model. Readers must therefore be aware that there are gaps and other issues that they have to take into consideration when planning and executing their own business models.

The research does not intend to explore how corporate environments of: cultures, values, norms systems and structures can affect shipbrokers' performance and motivation. It is not also intended to research how different cultures of shipowners' could change the perceived-value of shipbroking offerings. As such, the analysis focus is strictly limited to selected sample of drybulk shipowners' evaluation of the perceived value they receive from different strategic groups within shipbroking industry.

Design

To meet research objectives and answer its questions, I needed to develop a deductive hypotheses approach to identify the paradigm strategies, competed on market offerings and added-value services among the different shipbroking strategic groups and test how such attributes are perceived by selected sample of shipowners.

Cross-sectional survey strategy was applied to collect and analyze qualitative and quantitative data on the shipbroking offerings within different strategic groups. In addition, triangulation technique was used to test the findings with shipowners' perceived-value using case study strategy. The rationale behind applying cross-sectional survey strategy is to allow for large amount of data from sizable population to be collected and analyzed within limited time and budget. The case study approach on the other hand was used to find out the perceived value of various elements within real-life context and allow for critical evaluation of the existing paradigm among different shipbroking strategic groups.

Data was collected using multiple sources, including surveys, interviews and observation. The process was undertaken in two phases. The first phase used surveys targeting shipbrokers within the dry bulk industry as a sample to understand and identify the paradigm strategies, competed-on market offerings and other value attributes offered by different shipbroking strategic groups and the shipbroking industry as a whole. First, shipbrokers were asked to identify their strategic group on the basis of their marketing strategies. This included the following options:

- Single Segment
- Product Specialization
- Market Specialization
- Selective Specialization
- Full Market Coverage

Shipbrokers were then asked to identify the number of shipbrokers within their organisations, their covered industries, target markets and their core market offerings. They were then asked to rate the importance of each value attribute that they mostly invest in as an added-value service which best fit with their strategy. The criteria in which those attributes were chosen was mainly derived from Kotler's and Keller's (2010) value definition **(Value = Total Benefits – Total Cost)**. Table-1 illustrates the various value attributes offered by most shipbrokers regardless of their geographical location, covered markets, target customers and size.

Table-1

Customer Perceived Value			
Total Customer Benefit		**Total Customer Cost**	
Product Benefit	List of shipbroking core offerings	Monetary Cost	Total commission
Service Benefit	R&D, Market Research, Consultancy etc	Time cost	Total transaction time
Personnel Benefit	Personal Relationships	Energy cost	Breadth & width of services
Image Benefit	Brand Reputation	Psychological cost	Owners' perception on market offerings and shipbrokers' role.

The second phase of data collection carried out using case study approach. A questionnaire was prepared and distributed to major influencers in the shipowning organisation including CEO, CFO, Chartering Manager and Operations Manager. The questionnaire aimed to analyze the level of perceived value for each attribute offered by shipbrokers including the perceived value for different core market offerings. A combination of structured and semi-structured interviews was then conducted with the same sample to collect data on various motivational and psychological cues and buying behaviour that would affect their purchase decision. Most questions aimed to demonstrate how shipowners and ship operators perceive the contemporary shipbrokers' role and how such role meets with their changing needs and perceived value. Each interview took around 25 minutes and notes were taken future reference.

Methodology

All data was stored and mined in order to provide useful insight on the shipbroking industry and how each strategic group pursues its strategy. All data was then analyzed to show the differences and similarities within different strategic groups compared to the shipbroking industry as a whole and then test how their value attributes and market offerings are perceived by the selected sample of shipowners both as asset owners and as ship operators.

Various tables and charts were used to illustrate how such differences and similarities exist within the shipbroking industry as a whole and how the various attributes and market offerings are perceived by selected shipowners. In particular, the strategy canvas was used to analyze the different value attributes and services offered by the shipbroking industry and how such attributes are perceived by selected shipowners. The strategy canvas was introduced by Kim and Mauborgne (2005) as an excellent action framework for building a compelling blue ocean strategy. It was used to help us understand where the competition is currently investing, the factors the industry currently competes on and what customers receive from existing offerings. In the horizontal axis we show the various value-attributes which different strategic shipbroking groups compete on while the vertical axis shows the level in terms of importance or percentage of mostly invested in attributes. The selection of value attributes on the horizontal axis was mainly derived from the 'customer-perceived value' definition provided by Kotler and Keller (2010) as shown in table-1.

First, shipbrokers were segregated into strategic groups to identify differences and similarities within their strategies. These included three main factors which are: number of shipbrokers, covered markets and target customers. Then, a general strategy canvas was drawn to analyze the paradigm competed on attributes of market offerings and added value services. The results were then tested with the shipowners' perceived value for the same set of attributes where another strategy canvas was drawn to show the differences between the shipowners' perceived value in terms of asset owners and commercial operator. This allows for a better understanding of the industry and provides better insight on opportunities for value innovation through analyzing the gaps between the different value curves.

Sample Selection

Shipbrokers working within international dry bulk industry were chosen to be the first sample of the study. The main reason for choosing random shipbrokers from different markets is that shipping has become an international industry and most shipbroking services are offered to international markets and customers equally regardless of their geographical location.

Due to the explanatory nature of the research and the difficulty in providing sampling frames for shipbrokers; a purposive sampling strategy was used whereby heterogeneous sampling technique allowed collecting data and observing the differences and uniqueness among different shipbroking strategic groups and their value offerings. In addition, the research's credibility and validity was intended to depend on the data collection and analysis skills rather than the size of the sample.

Due to the limited resources and timeframe to conduct the research; data collection process could only take place during the month of August 2011. The shipbrokers' survey was sent by email to 400 contacts and a total of 46 surveys were filled (11.5% response rate). This was followed by further surveys conducted by telephone to ensure greater number of participation and ensure that brokers from all types of strategic groups participate. A total of 54 surveys were filled by shipbrokers. It is believed that such sample provided sufficient insight to be drawn to answer research questions and draw strategy canvas.

In contrast, a homogenous sampling technique was used when selecting shipowners' case in order to identify the perceived-value similarities and study them in great depth.

Here, I worked with one company that owns and operates a fleet of dry bulk vessels on international basis. For confidentiality reason, the company name cannot be revealed in this study. The selection of the company was based on the following reasons:

- Firstly, it has a reputable name and track history in the shipping industry and international dry bulk markets.
- Secondly, it consists of various business units and subsidiaries that separate shipowning from management and operations.
- Thirdly, it has performed extremely well over the past five years unlike most players in the industry.

My focus was to discuss, collect and compare data that are valuable for both shipowning and ship operating functions. It was made clear that separate success factors and value chain are required by the shipowner to handle their ship investment business unit compared to other units such as ship operations or ship management.

Research Limitation

In this research; I tended to deal with various reliability and validity threats. The different characters and expectations among different participants may have a level of participant error or participant bias which could affect the research reliability. Moreover, data collected from participants could be greatly influenced from an incident which was recently experienced, or it could be influenced by changes in the macroeconomic environment and thus may affect validity of data collected. Therefore, utmost care has been taken when analyzing data to ensure that data is reflected effectively and thus results can be generalisable.

Although a single case study has several limitations, particularly in its finding generalizations, it however proved to be a powerful tool to provide detailed and unique information that helped analyze the context and answer research questions. However, it is believed that a wider study on the shipbroking paradigm strategies, offerings and added-value services should provide a much deeper insight on opportunities for value innovation.

Results

Shipbrokers have been segregated into different strategic groups based on their marketing strategies. As shown in figures 3, 4 and 5; from all shipbrokers who participated in this research, we found that the shipbroking industry could be divided as follows:

- 11% follow a single segment strategy
- 19% follow product specialization strategy
- 15% follow market specialization strategy
- 50% follow selective specialization strategy, and
- 6% follow full market coverage

Figures 3, 4 and 5 illustrate how different shipbroking strategic groups follow different strategic and operational approach to handle their business in terms of number of shipbrokers (Figure-3), covered market (Figure-4) and target market (Figure-5). Figure-3 compares the number of shipbrokers working for each strategic group. Firms that have five or less brokers within their organisations are as follows: 50% from the single segment and product specialization groups, 75% from the market specialization and 56% from the selective specialization groups. Whereas, 75% of firms within the full market coverage group have more than 100 brokers.

By looking at figure-4, it could be seen that most of shipbrokers from all groups work within the freight market. 67% from the single specialization, 70% from the product specialization, 88% from the market specialization, 63% from the selective specialization and 100% from the full market coverage work within the freight market.

Similarly, when looking at figure-5, the majority of firms within all strategic groups except the full market coverage target charterers/cargo traders as main customers. This was represented by 50% from the market specialization and selective specialization groups while the product specialization group was equally split between targeting charterers/cargo traders and shipowners with 30% each. On the other hand, 67% from the full market coverage group target shipowners as main clients and 33% target charterers/cargo traders.

Figure-3

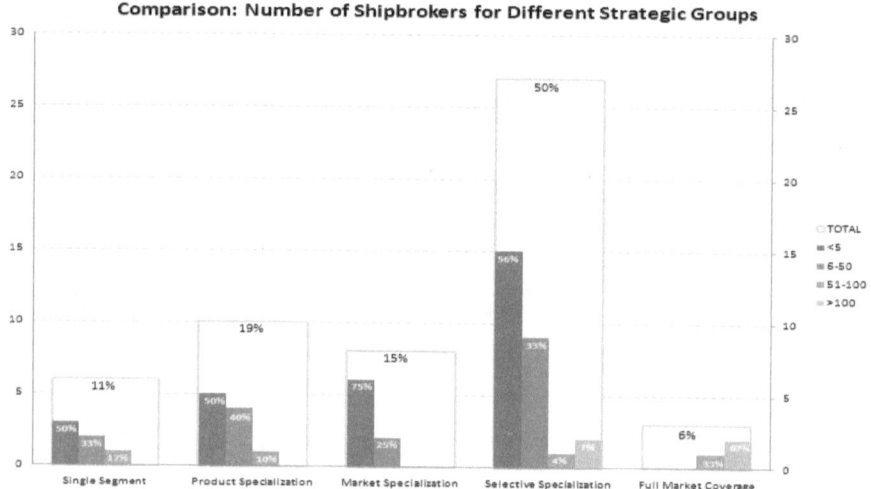

Figure-4

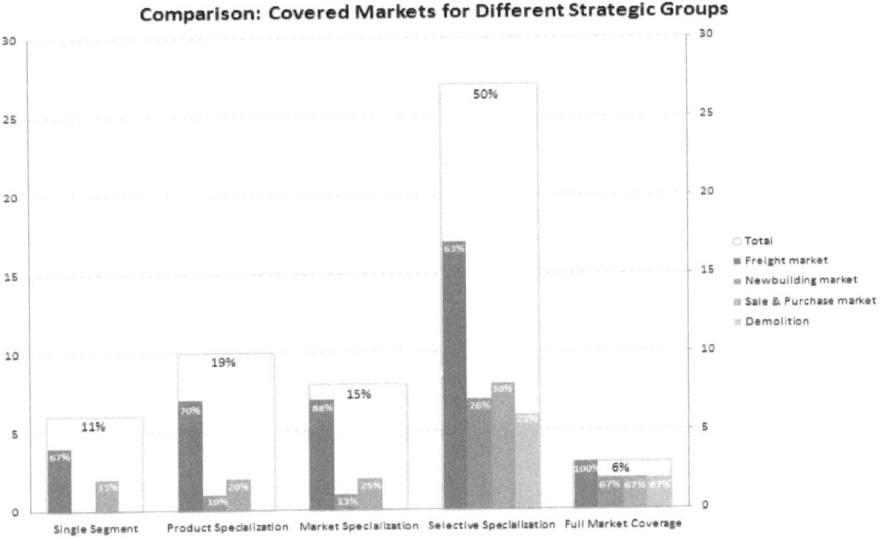

Figure-5

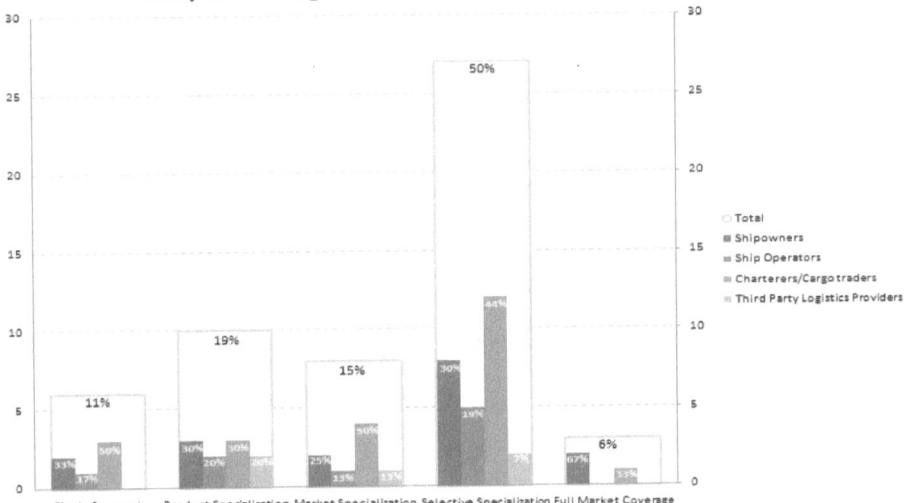

When considering the drybulk shipbroking industry as a whole, tables 2, 3 and 4 illustrate the differences between the overall paradigm strategies among industry players. Table-2 shows that 52% from all shipbrokers participated in this survey have less than five brokers within their organisations while the remaining were distributed as follows 31% have between 6-50 - 7% have between 51-100 and 9% have more than 100 brokers. Table-3 shows that 70% from all brokers work within the freight market, 33% within the demolition market, 30% within the sale and purchase and 20% within the newbuilding market. Table-4 shows that 43% from shipbrokers target charterers/cargo traders as main customers, 31% target shipowners, 17% target ship operators and 9% target third-party logistics providers.

Table-2 (Shipbroking paradigm strategy in terms of number of brokers)

Number of Shipbrokers	Count	Percentage
<5	28	52%
6-50	17	31%
51-100	4	7%
>100	5	9%

Table-3 (Shipbroking paradigm strategy in terms of covered markets)

Covered Markets	Count	Percentage
Freight market	38	70%
Newbuilding market	11	20%
Sale & Purchase market	16	30%
Demolition	18	33%

Table-4 (Shipbroking paradigm strategy in terms of target customers)

Target Customers	Count	Percentage
Shipowners	17	31%
Ship Operators	9	17%
Charterers/Cargo traders	23	43%
Third Party Logistics Providers	5	9%

Figure-6 illustrates the shipbroking paradigm competed-on market offerings. The value curve presents the overall percentage of shipbrokers offering each service. It shows that 67% from all shipbrokers who participated in this study offer voyage and time charter brokerage as part of their core-offerings, 41% offer ship agency services while only 7% offer fund management and investment brokerage. Table-5 provides detailed information on how different strategic groups concentrate on different market offerings and services as part of their strategies and provides the overall percentage from all participants on their core market offerings regardless of their strategic group.

Figure-6

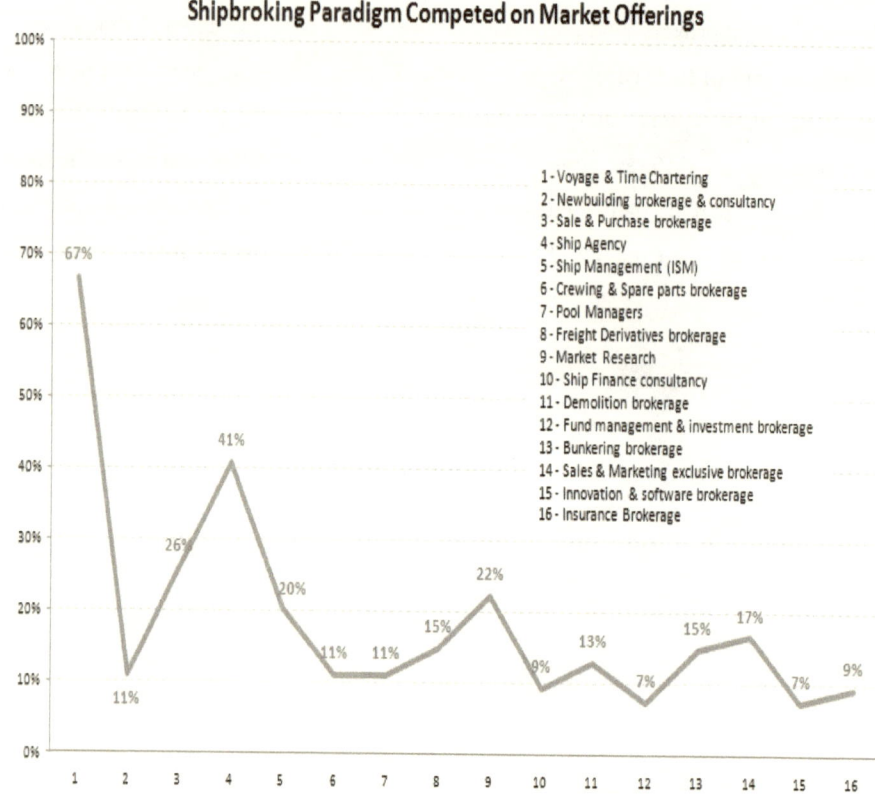

Table-5 (Percentage of shipbrokers offerings for different strategic groups vs. total average)

	Single Segment	Product Specialization	Market Specialization	Selective Specialization	Full Market Coverage	Overall %
Voyage & Time Chartering	40%	40%	88%	67%	100%	67%
Newbuilding brokerage & consultancy	10%	10%	0%	15%	33%	11%
Sale & Purchase brokerage	10%	10%	25%	33%	67%	26%
Ship Agency	10%	10%	63%	44%	67%	41%
Ship Management (ISM)	0%	0%	38%	26%	33%	20%
Crewing & Spare parts brokerage	0%	0%	25%	11%	33%	11%
Pool Managers	0%	0%	0%	19%	33%	11%
Freight Derivatives brokerage	20%	20%	13%	7%	100%	15%
Market Research	0%	0%	13%	30%	100%	22%
Ship Finance consultancy	10%	10%	13%	4%	67%	9%
Demolition brokerage	0%	0%	0%	19%	67%	13%
Fund management & investment brokerage	0%	0%	13%	4%	67%	7%
Bunkering brokerage	0%	0%	25%	15%	67%	15%
Sales & Marketing exclusive brokerage	0%	0%	0%	22%	100%	17%
Innovation & software brokerage	0%	0%	0%	11%	33%	7%
Insurance Brokerage	0%	0%	25%	4%	67%	9%

Other value attributes which were mainly derived from Table-1 included the breadth and width of services, personal relationships with customers, brand name, total commission, total transaction time, and market information and research. Figure-6 demonstrates the differences in value attributes offered by different strategic groups in terms of importance to each group's strategy. It shows six value curves, five of them represent different strategic groups while the sixth represents the weighted average value curve for the shipbroking industry as a whole. As shown in figure-7, the value curve for all strategic groups showed a similar paradigm. This means that the shipbroking industry is investing in same value attributes to support their strategies and deliver value to their customers. Although the level of importance varies from strategic group to another, but the overall trend could be clearly seen on where the competition currently invests in. There was high importance in the personal relationships and brand reputation as main factors to achieve this. In the contrary, the least invested in attributes were the breadth and width of services as well as the total transaction time.

Figure-7

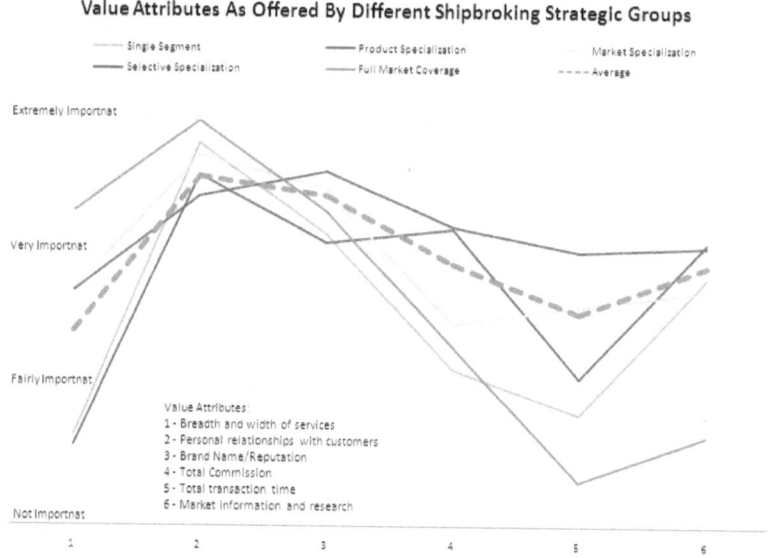

In order to identify the perceived value by shipowners on the different market offerings and value attributes offered by shipbrokers, a set of questionnaires were distributed to various influencers within the shipowner office including: CEO, CFO, Chartering Manager and Operations Manager. Participants were asked to answer the importance of each attribute in terms of perceived value for two separate business needs: as asset owner and as commercial ship operators. The average answers from each category were then calculated to present the shipowners' perceived value as shown in Figure-8 and Figure-9.

Figure-8

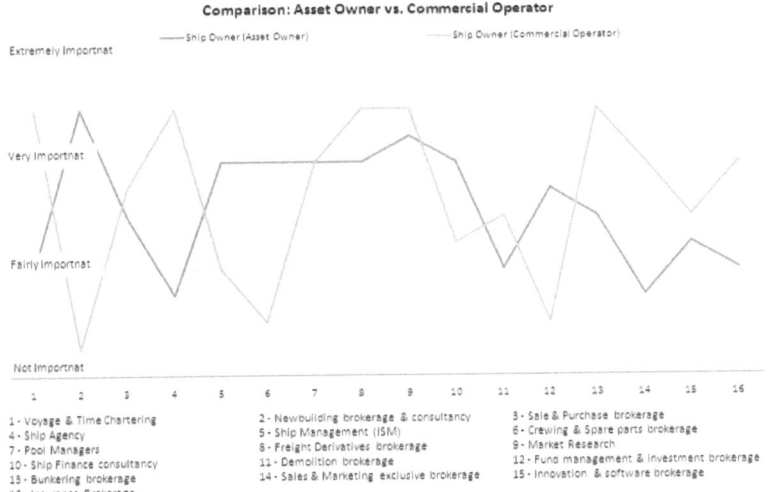

Figure-9

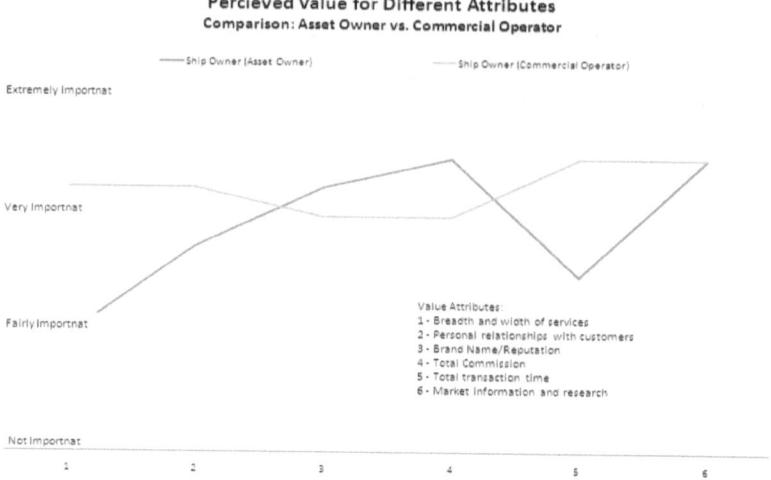

When considering how shipowners (asset owners) and ship operators (commercial operators) perceive the shipbroking market offerings (figure-8) and value attributes (figure-9), it was clear that each target customer requires a very different set of services and value attributes. Figure-8 shows that both shipowners and ship operators have almost opposite perceived value for most market offerings except for the pool management, freight derivates and market research services where a commonly higher perceived value were found. The ship operator for instance showed higher perceived value for chartering, ship agency, pool management, freight derivates, bunkering, insurance and market research. The asset owner on the other hand showed higher perceived value in newbuilding, sale and purchase, ship management, crewing, pool management, fund management and investment brokerage, freight derivatives and market research services.

In figure-10, the strategy canvas provides a comparison of three different value curves that present asset owners and commercial operators perceived value of various attributes

- as shown in figure-9 - and compare them with the average shipbroking mostly invested in value attributes as shown in figure-7. By comparing the value curve of paradigm shipbroking value attributes with the perceived value of shipowners and ship operators, we find that the level of differences is tremendous and that the value curve of shipbrokers is either over investing in some areas and under investing in others. The asset owner for instance showed higher perceived value for brand reputation, total commission and market research services while the ship operators seemed to require more of breadth and width of services, personal relations, transaction time and market research services.

Figure-10

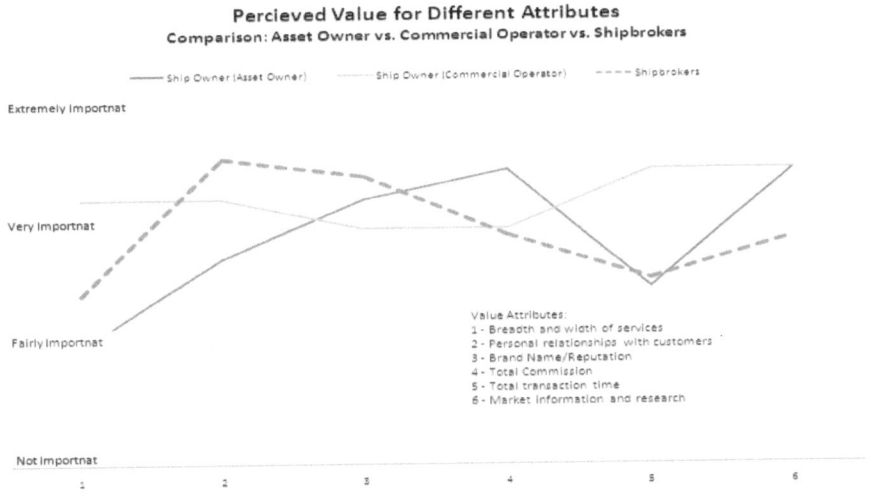

Similarly, when discussing the important role of modern shipbrokers with the ship operator, it was clear that the level of differentiation is mainly based on the level of market information and legal knowledge the broker can provide. This was found in the chartering manager's comment as follow: *"We consider our in-house shipbrokers a real asset, almost all dry bulk operators nowadays have their internal brokers; they seek, negotiate and finalize charterparties with huge efficiency, what we still lack is the*

systematic way to read the markets and take optimum decisions at the right time". In another statement drawn from the CFO on the importance of shipbrokers for their business requirements as asset owners, it was noted that their ultimate goal is to manage a solid financial operations with acceptable returns and cashflow. He stated that: *"when it comes to acquiring a second hand or a new ship, it is not the number of ships reported from the broker that makes us happy; neither is the price evaluation or personal advice but the set of customized proposals that make us meet with our business requirements and objectives more efficiently."*

Discussion

Shipbroking Paradigm Strategies

Firms within the dry bulk shipbroking industry work within different strategies. Depending on market attractiveness and firm's core-competences; shipbrokers decide to pursue the best strategy that could best utilize their resources and capabilities. In this research, shipbrokers were divided into five main strategic groups depending on their marketing strategies as follows:

- Single Segment Specialization
- Market Specialization
- Product Specialization
- Selective Specialization
- Full Market Coverage

Among these groups, our analysis shows that there are various strategic differences and similarities found mainly within covered markets, target segments and the number of brokers needed to implement such strategy. Other categories include the breadth and width of services, market offerings and added-value services.

The results show that most shipbrokers within most strategic groups have similar paradigm in terms of markets covered, target segments, market offerings and added-value services. They even have similar strategies in terms of number of brokers needed to implement such strategies. Our analysis show that the majority from shipbrokers work within the freight market (70%) targeting charterers/cargo traders as main customer (43%) by offering them voyage and time chartering services (67%) with five or less brokers within their organisations (52%). The analysis also show that most shipbrokers concentrate on personal relations and brand reputation as main added-value services to support their strategies and increase value proposition.

The fact that 50% of shipbrokers who participated in this study pursue a selective specialization strategy could mean that they do not see in one specialized segment or particular product a profitable business or growth opportunities and thus aim to continuously seek for other attractive segments within the industry and create new offerings for each segment which may further increase competition on the international level. It could be argued that as most industry players aim to do the same thing and compete over the same customers without having specialized and fully aligned resources and capabilities, this will most probably lead the industry into a decline stage or be threatened by new specialized entrants or substitute industries.

For instance, while most shipbrokers target charterers/cargo traders within the freight market by offering voyage and time chartering services, the rise of specialized third, fourth and fifth party logistics providers in recent years focusing their offerings on specific industries and delivering fully integrated supply chain solutions with improved lead-times could bring further challenge for shipbrokers targeting this segment. In addition, the commoditization of the chartering service itself could also raise the question whether charterers and shipowners will continue working through brokers or that fixing direct between principals will become a norm?

Shipbrokers must therefore aim to reconsider their strategies in terms of niche-segments, markets and competences in order to differentiate their position among other rivals. Dr. Lorange (2009) divided the shipping industry into four specialized organisations each focusing on different success factors (Owning Steel – Operating Steel – Using Steel – Innovating around Steel). He claims that in order for firms to deliver superior value, they must concentrate on one or two competencies and outsource the non-core. The buying behaviour and needs for each type will therefore differ. Accordingly, shipbrokers who are targeting charterers, shipowners or ship operators must understand this fact and design market offerings and value attributes enabling them to deliver increased perceived value to their target customers.

The analysis shows that shipowners have different needs and perceived value than those of ship operators', in fact - except for market research, freight derivate and pool management services with commonly high perceived value - the analysis shows that they have almost opposite perceived value for most other market offerings. The ship operator for instance seeks to employ a ship in the most efficient and productive way. They showed higher perceived value for chartering, ship agency, bunkering and insurance brokerage. In terms of added-value services, the ship operator showed higher perceived value for total transaction time and market information. The asset owner on the other hand showed higher perceived value in newbuilding, sale and purchase, ship management, crewing, fund management and investment brokerage combined with lower cost and customized market information and research.

When comparing perceived value of shipowners and ship operators to what is offered by shipbrokers; it was evident that there are huge gaps in many areas including the shipbroking paradigm market offerings and added-value services. This means that the dry bulk shipbroking industry is either over or under investing in many value attributes that do not collectively provide increased perceived value to any specific segment.

My claim is that – being a central part of the shipping industry for hundreds of years, shipbrokers must find new ways to meet the changing demand and bridge these gaps to sustain a strategic role within their supply networks. This may require redefining their strategic role, the business they are in, the customers they serve and what services to offer them and the way such services are offered (Markedis 2008). Instead of pursuing traditional strategies of trade-offs between cost and service, shipbrokers are advised to achieve both simultaneously through the business model innovation approach. Forward-looking shipbrokers that can understand such trend and develop strategic intent to create their own business models will be able to shift away from commoditization and enjoy the first move advantage.

Business Model Innovation – A source of competitive advantage

In this paper, a general analysis of industry value curve was drawn. It highlighted the most competed-on attributes and how such attributes were perceived by our selected sample of shipowners. It was evident in this paper that there are different needs and perceived value by the different market segments; fairly obviously, players within each segment will also have different strategies and thus different perceived value.

While the modern drybulk shipowners are increasingly dealing with much more complex environments within the four shipping markets, the different tasks required for managing their value chains have also become more sophisticated. There could be however some common jobs-to-be-done among wider sample of shipowners that shipbrokers could build on these commonalities and create higher demand and profitability. As most shipbrokers seek to differentiate their offerings through finer segmentation, optimized processes and refined metrics, the intensive head-to-head competition and the dynamic buying behaviour in such highly mature, volatile and fragmented industry is driving profitability and growth opportunities down to a point where sustainability could be threatened. Therefore, in order to challenge such prevailing economic model and change the rules of the game, shipbrokers must develop innovative strategies that open up new markets, sideline competition, and create higher value not only for shipowners but for wider market segments as well as other internal and external stakeholders.

The importance for shipbrokers to pursue a business model innovation approach is of vital importance to increase their customer value proposition and maintain a sustainable role within their supply networks. It could be argued that many of the dry bulk shipbrokers are cautiously or incautiously work within their own business models. However, what is found in this research is that most of those business models are similar in terms of customer value proposition, profit formula, key resources and key processes.

It was also found that most shipbrokers follow similar strategy in terms products and services offered, targeted segments, and the kind of activities performed.

Johnson (2010) provided the excellent four-box framework which if innovatively designed and integrated; shipbrokers will be able to develop an innovative business model: customer value proposition, profit formula, key resources and key processes. He argued that for firms to generate better insight on value proposition, they should think about the four common barriers from keeping customers getting their job done: insufficient wealth, access, skill and time (Johnson 2010). Fairly obviously, creating new business models is neither a straightforward nor an easy task. With it come many opportunities and challenges which companies must manage and be aware of. Markides (2008) claims that firms can only create innovative business model by being able to innovatively redefine the business, customers, services and the way such services are provided. This means that business model innovators should constantly create a new what, a new who and a new how to be able to change the rules of the game.

By using the business model innovation approach, shipbrokers could gain a unique way of thinking which would help them achieve value innovation and bridge the gap between the corporate, business and operational strategies. The business model innovation could be used as an excellent tool for analysing strategic choices and developing systematic approach focusing on all the elements of the organisation including commercial, financial and operational structures. The result is an innovative customer value proposition that meets with customers' perceived value. By eliminating, reducing, creating and adding value attributes, shipbrokers will be able to create a leap in their value curve which would change the rules of the game and create new market place and higher profitability (Kim and Mauborgne 2005).

This means that business model innovators should be able to offer different value propositions to different set of customers while having different internal processes, systems, cultures and structures compared to what other rivals have. Accordingly, shipbrokers who aim to pursue business model innovation approach will have different key success factors compared to other rivals with locked-in strategies and similar paradigm competed-on attributes (Markides 2008).

Consistent with the resource-based-view of the firm, the business model innovation is therefore considered a source of sustainable competitive advantage that would be difficult for other players to imitate, substitute or replicate quickly.

A New Shipbroking Role

Our research shows that the current shipbroking role is an intermediary between shipowners and charterers or the buyers and sellers of ships. The role of intermediaries has been there since the 18th century when shipbrokers started to help shipowners to employ, sell or buy their ships by linking them with the international markets. While this assumes that shipbrokers should only work within the freight and sale and purchase, Dr. Martin Stopford (2008) identified four closely related shipping markets that collectively form how the industry operate. Thus, it would be imperative to include all four shipping markets when describing the shipbroker role which in my opinion does not only reflect the actual role but also opens up other opportunities when executed properly.

In addition, the advent of various technological innovations and the diffused chartering knowledge has changed the perceived value by shipowners and ship operators. Many functions which used to need specialized companies could now be handled by one or two brokers at shipowners' office. Although the shipbroking profession has evolved through many stages and cycles responding to various economical, technological and social changes, there is a clear trend that the modern shipbrokers' role has been shifted from 'Agents' to 'International Intermediaries' to 'information intermediaries' to become 'execution only intermediaries'.

Today, as most markets have shifted to buyers', a clear trend towards specialization is seen in most industries including the dry bulk shipping (Kotler & Keller 2010). When considering the dry bulk industry value chain; there is an empirical proof that the requirements of shipowners differ from those of ship operators' or those of charterers'. Each player within the supply chain has different success factors and perceived value.

Therefore shipbrokers' should not only help their customers find an employment of their vessels or buy and sell vessels but also enhance their capacity to do so. They must understand and integrate with their customers' value chains (Porter 85). They must help them innovate new offerings, optimize their performance and adapt to various types of changes (Kotler & Keller 2010). In order to achieve this, shipbrokers need to consider another transformational shift away from their existing role "execution only intermediary" and aim for a more sustainable role which I would call "innovation intermediary".

In my opinion, innovation intermediaries should aim to enhance their customers' performance by managing, linking and bridging an innovative network of stakeholders in an optimized and sustainable way. They should deliver innovative value to customers by integrating many of their support and primary organisational functions with other industry stakeholders such as providing customized research services, financial management, marketing implementation, HR outsourcing etc. I would argue that their primary role should include the vertical and horizontal integration of their customers' value chain with industry stakeholders to not only add value through quality of service and increased earnings, but also in adding value through all parts of their customers' organisations and business needs.

In sum, my recommendation for a new role of shipbrokers could be broadly defined as: **"Innovation intermediary that provides vertical and horizontal value chain integration with a network of stakeholders to enhance the performance of the ship** in terms of operations, management or investments in an optimized and sustainable way".

Conclusion & Recommendations

Shipbrokers working within the dry bulk industry are facing strong head-to-head competition. Being in a highly mature industry, most rivals are trying to differentiate their offerings by increasingly investing in different attributes such as brand reputation, diversification, technology, R&D budgets as well as many others. The paper shows that most of the shipbroking firms have similar strategies, paradigms and value curves.

In the past two decades, the technological development and diffused knowledge have invited many other players such as financial institutions into the industry to offer more specialized services. At the same time, shipowners are working with much more complex environment compared to how they used to work in the past. Today, shipowners have to deal with various international stakeholders combined with increasing challenges of volatility and uncertainty. In addition, it was evident in this paper that there are gaps between the perceived value of shipowners and what is offered by shipbrokers. This means that shipbrokers are either over or under investing in different attributes which may or may not be valuable to shipowners. On the medium to long term this might affect their growth capability and lose their role as intermediaries within their industries.

In order to sustain and improve their strategic role, shipbrokers need to reconsider their strategies and position themselves as vital players within international supply chains. This requires redefining the current shipbroker role to a more innovative and sustainable one. The new role must open new areas for brokers to offer innovative customer value propositions and market offerings with higher customer perceived value. In this paper, I recommended a new role called 'innovative intermediary'. Shipbrokers need to constantly create customer value propositions by helping their customers to better perform the different functions within their value chains. In order to do this, the business model innovation approach is recommended as not only a tool of analysis to create new value for bigger markets but also to deliver such value in the most efficient and effective way.

As shipbrokers managed to evolve through many lifecycles and stages, it is of vital importance for policy makers, international bodies as well as practitioners to reconsider the paradigm shift that is not only changing the shipping industry but almost all industries around the world and try to develop new practices and methods for the shipbroking profession which shall ensure its long term sustainability.

References

- Martin Stopford (2008), Maritime Economics (3rd edition), Routledge.

- Robert M. Grant (2008), Contemporary Strategy Analysis (6th edition), Wiley

- Kotler, Keller, Brady, Goodman and Hansen. (2009). Marketing Management (1st European Edition), Pearson.

- Kim, W.C and R. Mauborgne (2005). (1st Edition). Blue Ocean Strategy: How to create uncontested market space ad make the competition irrelevant. Boston, Harvard Business School press.

- Mark W. Johnson (2010), (1st Edition), Seizing the White Space: Business Model Innovation for Growth and Renewal. Harvard Business Press.

- Peter Lorange (2009). (1st Edition). Shipping Strategy: Innovating for Success. Cambridge University Press

- Constantinos C. Markides (2008). Game-Changing Strategies: How to create new market space in established industries by breaking the rules. London Business School. Jossey-Bass

- Michael E. Porter (1985), (1st Edition), Competitive Advantage: Creating and sustaining superior performance. Free Press

- Alan E. Branch (1996), (7th Edition), Elements of shipping. Chapman & Hall

- Joan Magretta (2002), Why Business Models Matter? Harvard Business Review (May 2002): 87-92

- J.B Barney (1991). Firms resources and sustained competitive advantage. Journal of management 17 (1991): 99-120

- International Maritime Organization IMO, (2010). (Online) available from http://www.imo.org (accessed on 10.7.2010).

- United Nations Conference on Trade and Development UNCTAD, (2009). Review of maritime studies (online) available from http://www.unctad.org/en/docs/rmt2009_en.pdf (accessed on 27.9.2010)

- Baltic International Maritime Organisation BIMCO (2010). (Online) Available from www.bimco.org (accessed on 27.11.2010)

- Krishna Prasad (2008). Changing Role of Shipbrokers: A study of the impact of modern communication in practical ship-broking. (Online) Available from http://www.mu.ac.in/arts/social_science/economics/p16.pdf (Accessed on 20.7.2011)

- The Baltic Exchange, (Online). available from www.balticexchange.com (accessed on 27.11.2010)

- The maritime advocate (2011). Adding value to shipbroking solutions. (Online) available from http://www.maritimeadvocate.com/developments/adding_value_to_shipbroking_solutions.htm (accessed on 20.08.2011)

- Sustainable shipping initiatives – the case for action (2011). (Online) available from www.forumforthefuture.org (accessed on 19.6.2011)

- C.K Prahalad and Gary Hamel (1990). The core competence of the corporation. Harvard Business Review (May-June 1990): 79-91

- Mark W. Johnson, Clayton M. Christensen and Henning Kagermann (2008). Reinventing your business model. Harvard Business Review (December 2008): 51-59

- Christensen, C., and Raynor, M. (2003). The innovator's solution. Boston: Harvard Business School Press.

- The virtual shipbroker (2009). Shipbroker Fast Track. (Online) Available from http://www.virtualshipbroker.com (Accessed on 04.05.2010)

- The role between a broker and the buyer and seller of a ship is key but it can sometimes be problematic, not least because there is no Act of parliament governing it. (2006). English law has trouble finding definition of a shipbroker. (Lloyd's list). (Online) available from http://www.lloydslist.com/ll/sector/regulation/article76931.ece (accessed on 16.6.2011)

- Lewis R. Fischer (1993). A bridge across the water: Liverpool shipbrokers and the transfer of Eastern Canadian sailing vessels, 1855-1880. The Northern Mariner/Le Marin du nord, m, No. 3 (July 1993): 49-59.

- Lehmann-Ortega and Marc Schoettl. (2005). The role of business models in strategic innovation. Article presented at CLADEA, Santiago de Chile, October 2005

i want morebooks!

Buy your books fast and straightforward online - at one of world's fastest growing online book stores! Environmentally sound due to Print-on-Demand technologies.

Buy your books online at
www.get-morebooks.com

Kaufen Sie Ihre Bücher schnell und unkompliziert online – auf einer der am schnellsten wachsenden Buchhandelsplattformen weltweit! Dank Print-On-Demand umwelt- und ressourcenschonend produziert.

Bücher schneller online kaufen
www.morebooks.de

 VDM Verlagsservicegesellschaft mbH
Heinrich-Böcking-Str. 6-8   Telefon: +49 681 3720 174   info@vdm-vsg.de
D - 66121 Saarbrücken   Telefax: +49 681 3720 1749   www.vdm-vsg.de

www.ingramcontent.com/pod-product-compliance
Lightning Source LLC
Chambersburg PA
CBHW031548210526
45464CB00003B/1211